Copyright © 2016 by Hugh Morrison

All rights reserved. This book or any portion thereof may not be reproduced or used in any manner whatsoever without the express written permission of the publisher or author except for the brief quotations in a book review.

www.struartapp.com

First Edition 2016
Second Edition 2017

ISBN 978-1-78222-479-2

Production management by Into Print
www.intoprint.com
+44 (0)1604 832149

Printed and bound by UK and USA by Lighting Source

Introduction

Exercise Book

Pocket Book Companion to

Structural Engineering Art and Approximation

Student and CPD Exercises

This book is an *accompaniment to the book:* 'Structural Engineering Art and Approximation'; which is currently in its 3rd Edition (May 2016).

Feedback received has made clear that some of the arguments and presentations in 'Structural Engineering Art and Approximation' are a little haphazard and not particularly easy to follow if the subject matter is studied without reference to specific problems. The exercises are a remedy to this, for those who wish to follow the book through; taking the reader through the book to aid familiarity with the content assisting future reference to the book. The questions are further intended to demonstrate the design process in an uncomplicated way; related to realistic problems.

An annotated book copy may also be kept and used as a record of continuing professional development (CPD) for those engaged in the profession.

There have been complimentary reviews but little feedback on the subject matter. Whilst the book is very much an idiosyncratic approach to the subject, the message behind it is important. Engineers need to learn during their training and education methods of evaluating designs which preclude excessive computer use. These methods should be simple; backed up by sketches and hand calculations.

The reader is encouraged to work through all of the exercises, which will, if undertaken with reference to the sister publication, *Structural Engineering Art and Approximation*, follow the principal chapter topics from front to back. Blank pages are included to enable notes to be taken, whether on a tablet or traditional hard copy. It is suggested to work through the problems on a separate piece of paper and to transfer useful notes for future reference in the exercise book.

Exercise Book

The answers may be found from three main sources:

1. Rear Section of Book

In the rear of the book (see contents). Answers are summarised and methods for obtaining the answers are referenced to relevant sections of the book.

2. YouTube Videos

For each question hand calculations with narration follows in more detail the calculations made. See http://struartapp.com/videos/

3. Video Transcripts

For each question hand calculation transcripts may be downloaded from the website in the resources area. See http://struartapp.com/

Exercise Book

What may be Achieved by Undertaking the Exercises

It is highly recommended that the reader takes the time to work through the problems. This might take a day or more. With cross referencing to the book - '*Structural Engineering Art and Approximation*' - it is hoped that the initiate to professional practice or university student will have gained:

1. **Appreciation of the Design Process**. *Real examples are selected to illustrate areas of interest during concept design stages; noting that design solutions are never exact; there are always other approaches which may give slightly different answers; there is seldom a 'cut and dried' answer.*

2. **Enhanced Understanding of Simple Structures** . *Simply supported beams, cantilevers and continuous beams etc. which may be used as part of the practising engineer's 'toolbox'.*

3. **Confidence in Conceptual Design Methods**. *With visual, and simple guides to selecting appropriate structural elements.*

4. **Addition to a Design 'Toolbox'**.*Engineers in practice will generally build up a portfolio, or 'toolbox' of useful simplified methods to inform future designs suited to that individual.*

5. **Knowledge of the Book: *Structural Engineering Art and Approximation***: *Knowing where to look in the book for future reference and a better appreciation of the subject matter and uses of the book.*

6. **CPD points**. *'Continuing Professional Development'- a good 8 hours or more of CPD with the blank pages to record progress/thoughts/evidence.*

Any feedback would be appreciated. Any advice on improving the videos, for instance, would be accepted with gratitude. Contact may be made through the website http://struartapp.com/contact/

Contents

Exercise Book

Question 1

Simply Supported Beam

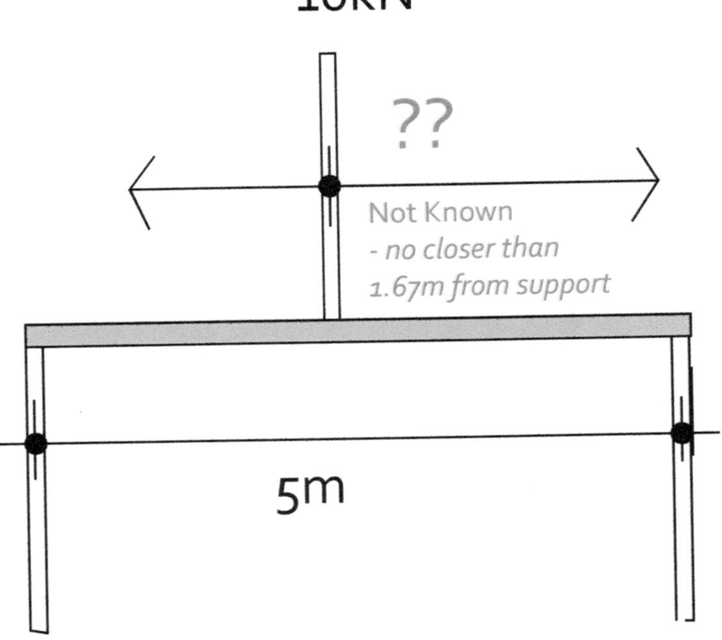

Beam 178x102x19UB
$I=1360\ cm^4$ $E=205\times10^3\ N/mm^2$ (mPa)

Hint ....
- Fig. 2.2.13

Background

In the first floor of a house design a beam is needed to support the roof. It is not known where the post supporting the roof is located along the beam span - except that it will be no closer to the support than 1.67m (third span). The contractor demands that the beam is sized for ordering materials.

Simply Supported Beam

Questions

1(a)

Find the maximum design bending moment based on the variable loading. Ignore any ultimate load factors for this exercise.

1(b)

Find the maximum design reaction on the posts.

1(c)

What is the worst case deflection in mm under load (in varying position) shown?

1(d)

Is the deflection of the beam for the size selected considered to be excessive; or is it acceptable for normal domestic purposes?

Hints

- *See Figure 2.2.13 for comparison of bending moments and shears for varying load position.*
- *Equations to use for bending, shear and deflection may be found within chapter 2.2 sufficient for design.*
- *Expressing a ratio of the span/deflection is a standard way to assess the magnitude of the deformation.*

Question 1

Simply Supported Beam

Question 1

Simply Supported Beam

Question 2

Simply Supported Beam

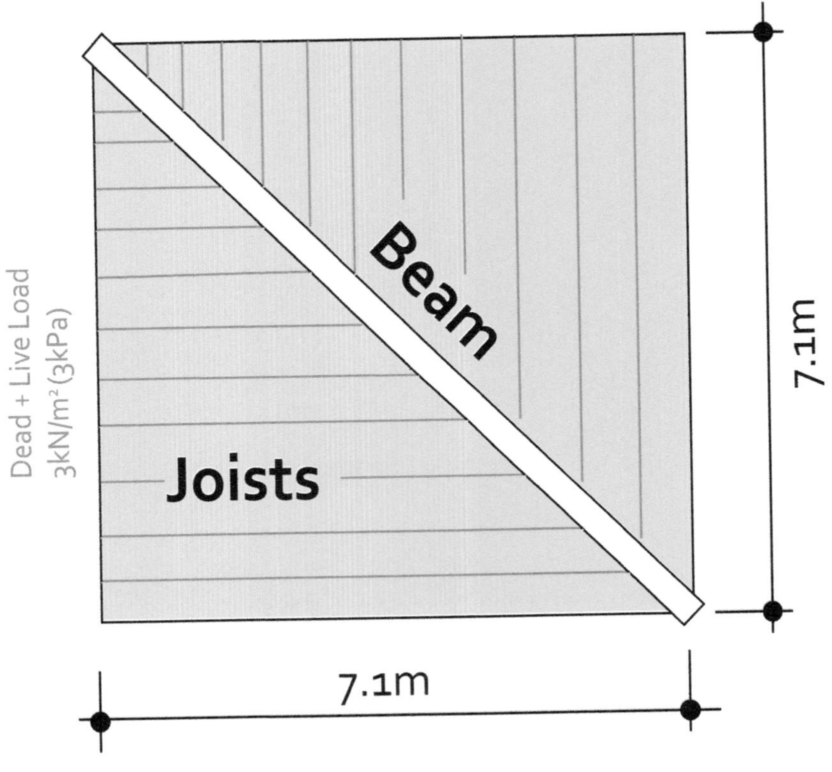

Beam 533x210x101UB
I=61520 cm^4 E=205x10^3 N/mm^2 (mPa)

Hint ....
- Fig. 2.2.8

Background

In an apartment block one of the accommodation floors at the corner of the building is set out diagonally across a 7.1x7.1m space. The joists span orthogonally to the main building grid. The beam size selected is shown with relevant properties.

Simply Supported Beam

Questions

2(a)

Work out the beam span and sketch the loading pattern on an elevation with simple supports. Evaluate peak load as units kN/m.

2(b)

Using an equivalent uniformly distributed load (UDL) and the relevant formulae in chapter 2.2, calculate the maximum estimated bending moment.

2(c)

Find (for the loading in 1(b) assumed) the estimated beam deflection for the chosen section in mm.

2(d)

Calculate the maximum support shear for both of the loading cases above (UDL and triangular loading).

Hints

- *See Figure 2.2.8 for loading substitution.*
- *Equations to use for bending, shear and deflection may be found within chapter 2.2 sufficient for design.*

Question 2

Simply Supported Beam

Question 2

Simply Supported Beam

Question 3

Cantilever Beam and Support Post

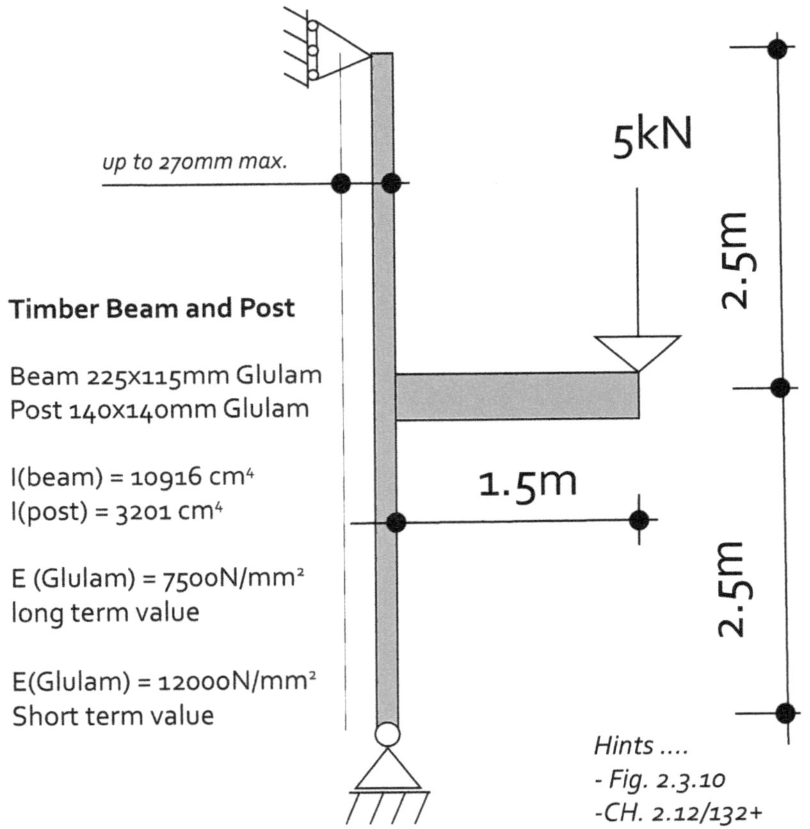

Background

A client in a commercial development in timber frame construction wishes to have exposed glulam balcony framing on the front elevation. The balcony beam may not be any larger than the size shown as discussed with the architect, but the post may be deepened in section (but not widened).

Cantilever Beam and Support Post

Questions

3(a)

Find the deflection of the beam assuming a pure cantilever section. *Use long term modulus E for this calculation*. Sketch bending moment diagram and deformed shape in beam and post.

3(b)

Find the deflection of the beam tip, ignoring its own flexure, based on the deformation of the post. The post is restrained horizontally at the top and pinned at the base. *Use long term modulus E for this calculation.*

3(c)

Find total deflection [3(a)+3(b)], long term) and approximate fundamental natural frequency, *fn*, at the beam tip (*use short term modulus E*). *Assume for calculating fn load is dead + 10% live load*. Is this deflection acceptable?

3(d)

If the post is increased to a 270x140wide section - 270mm into the building as shown by the dotted line - find the *total* deflection long term and approximate fundamental natural frequency as above. Is this deflection acceptable for normal domestic purposes?

Hints

- *See Figure 2.3.10 and 2.3.11 for examples of post support.*
- *Figure 2.3.2 for cantilever deflection and bending moment.*
- *Chapter 2.9/105 for rotation of beam/post with moment at end.*
- *Watch dynamics (i.e. liveliness of beam). See Ch. 2.12 / 131-132 calculation and table to work out approximate natural frequency based on tip deflection.*

Question 3

Cantilever Beam and Support Post

Question 3

Cantilever Beam and Support Post

Question 4

Continuous Beam

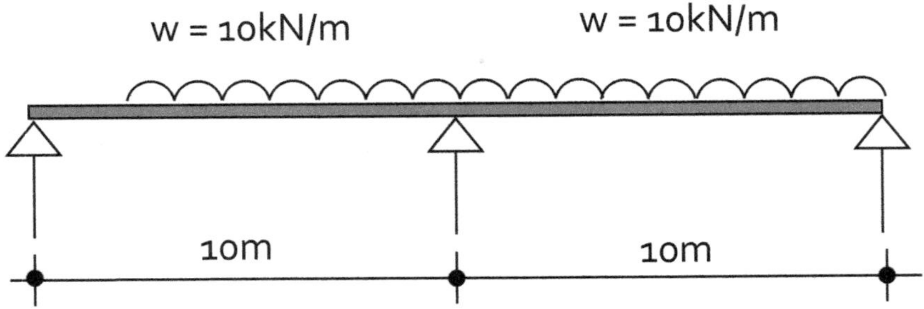

Beams and posts 533x210x109UB
I=66820 cm^4 E=205x10^3 N/mm^2 (mPa)

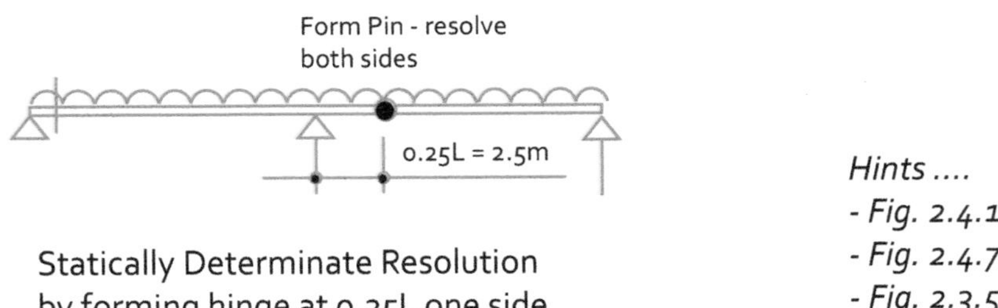

Statically Determinate Resolution by forming hinge at 0.25L one side

Hints ....
- *Fig. 2.4.1*
- *Fig. 2.4.7*
- *Fig. 2.3.5*
- *Fig 2.3.7*

Background

In an office building with 10mx10m grid the primary beams have been selected as the size noted above. There is a transportation problem in that any beam greater than 13m in length cannot be delivered without a special escort vehicle at prohibitive cost. A solution is requested preferably avoiding heavy moment connections (thick plates and bolts) - by forming a hinge.

Continuous Beam

Questions

4(a)

Forming a hinge point at a distance of 25% span from the central support calculate the support moment in the left span and maximum span moment in the right hand span. Find the support reaction in the middle support.

4(b)

By breaking the beam down into components by splitting either side of the hinge and inserting equilibrating shears at the hinge, calculate the approximate maximum span deflection in one of the beams.

4(c)

Check the expected bending moments and deflections for a continuous beam using equations provided in the book (two equal spans UDL both spans). How do the results compare with 4(a) and 4(b)?

Hints

- *See Figure 2.4.1 for standard 2 span beam with UDL both spans*
- *Figure 2.4.7 for hinged beam results.*
- *Figure 2.3.5 and 2.3.7 for a method of working out tip deflection in the left hand side split beam by pro-rata factoring against fixed end cantilever.*

Question 4

Continuous Beam

Question 4

Continuous Beam

Question 5

Three Storey Wind Frame

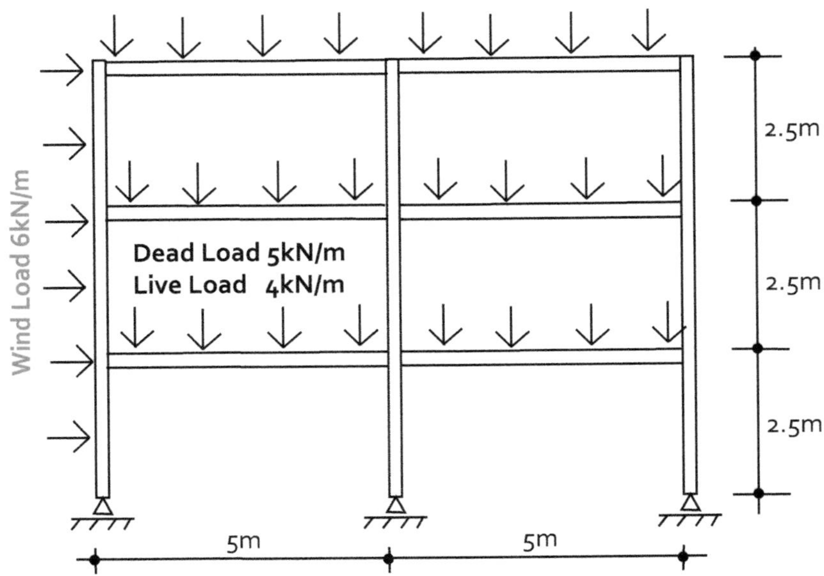

Timber Floor with Steel Beams

Beams and posts 254x146x31UB
I=4413 cm^4 E=205x10^3 N/mm^2 (mPa)

Plastic Modulus (S) = 393 cm^3

Hints ....
- *Fig. CH. 2.5/43-44*
- *Fig. 2.2.3*
- *CH. 2.12/132-133*
- *Ch. 2.12/138-139*

Background

A three storey office building is to be designed without using cross bracing as the architect would like fully glazed façades. The building has timber floors. Frames are 5m spans. There are four frames of identical composition. Loading has been given on the diagram for wind, dead and live. All floors have the same loading.

Three Storey Wind Frame

Questions

5(a)

Check the beam size *for bending strength only* assuming fully restrained *laterally* along its length by the joists and simply supported at each column. The section properties are outlined in the question. For simplicity use a safety factor of 1.5 for ultimate loading (the sum total of dead and live loads).

5(b)

By applying virtual pins at upper storey columns and in the beams sketch wind loading bending moment diagram. Calculate the approximate bending moments in the posts *at top of ground floor only* from wind loading; without factoring the load.

5(c)

Find the approximate sideways/lateral deflection under wind load at each storey (e.g. storey drift as net lateral deflection per floor) without resorting to a computer model. Sum these to give the total building wind deflection.

5(d)

Taking dead and 10% of live load work out the deflection in the beams assuming simple supports. Using an empirical formula work out the fundamental natural frequency of the beams. Is this likely to be a potential problem to the occupants in terms of liveliness of the floors?

Hints

- *See Chapter 2.5/43-44 for wind frame method.*
- *Figure 2.2.3 simply supported beam.*
- *Chapter 2.12/132-133 and 2.12/138-139 for dynamic checks.*

Question 5

Three Storey Wind Frame

Question 5

Three Storey Wind Frame

Question 6

Warren Truss

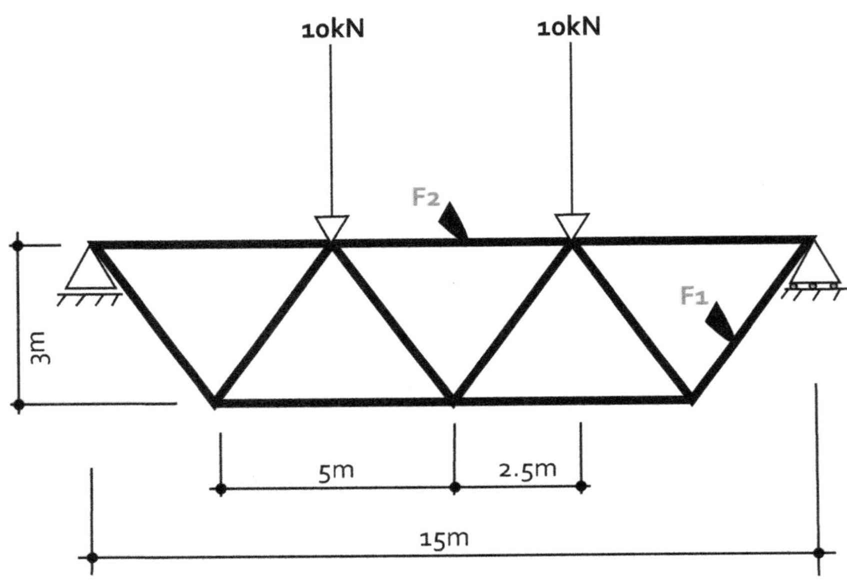

Truss Struts all 76.1x3mm CHS
(circular hollow section)
STEEL
E=205x10³ N/mm²; G=78x10³ N/mm²

Area of section = 6.9cm²

Hints ....
- Fig. 2.2.9
- CH. 2.6/53-55

Background

A glazed atrium roof in a prestigious office building comprises welded tubular steel sections. Two of the trusses support the feet of a raised roof-light feature which are 1 tonne or 10kN each. .

Question 6

Warren Truss

Questions

6(a)

Calculate **F1** - the first internal strut force - by resolving at the support. State if tension or compression.

6(b)

Calculate **F2** - the top chord force - by resolving for the midspan moment in the truss. State if tension or compression.

6(c)

Check the axial stresses in the struts (76.1x3CHS section) using the area provided. Multiply the load by 1.50 to represent ultimate load factor for the check. Is buckling likely to be an issue? *Assume all joints laterally restrained.*

6(d)

Calculate the deflection based on flexure and shear using the approximate methods employed in chapter 2.6. *Sections are steel* - see diagram.

Do not use the factored load as this is a serviceability condition.

Hints

- *The quickest way to size the truss elements is to resolve at the support and at midspan for top and bottom chords.*
- *See Figure 2.2.9 for internal forces and reactions.*
- *Chapter 2.6/53-55 for method.*
- *Chapter 2.7/78 for Euler buckling check.*

Question 6

Warren Truss

Question 6

Warren Truss

Question 7

Warren Truss

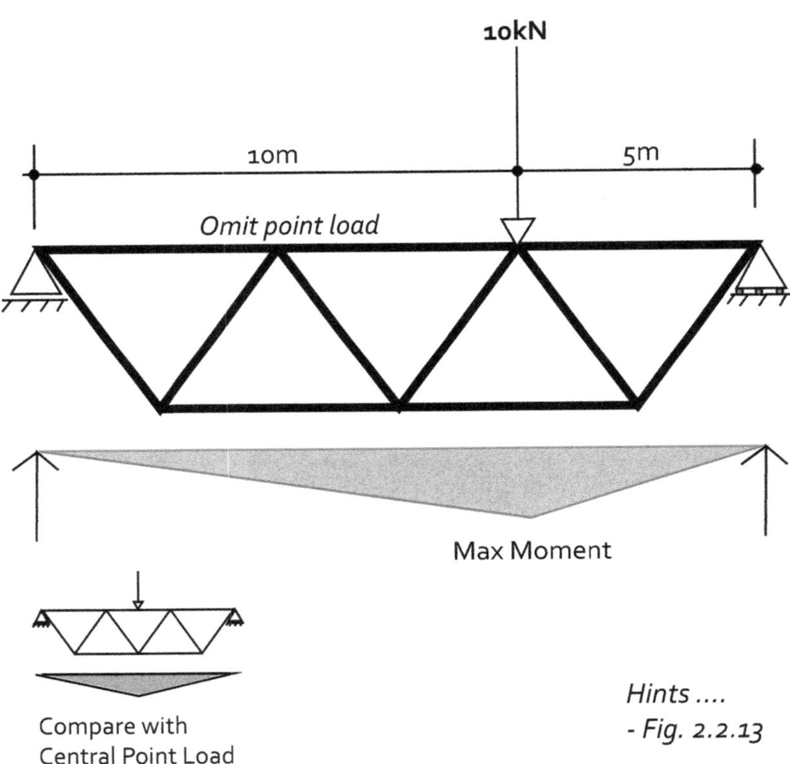

Background

On of the trusses in the glazed atrium roof has an undecided location for a window cleaning cradle support (total weight 1 tonne). The point load from the cradle is to be located either in the middle of the struts or at the internal top node (shown above).

Question 7

Warren Truss

Questions

7(a)

Calculate overall frame bending moments and shears for two conditions: (i) central point load (ii) point load as located on the diagram.

7(b)

If the 10kN load was in the centre of the span what would be the bending moment in *the structure overall* at midspan? What percentage of this moment is the bending moment calculated for the beam in the diagram - *ref. 7(a).*?

7(c) For the loading position in the diagram [ref. 7(a)] what is the worst case force in the first internal strut (which was calculated in the previous example as **F1**)?

What is the value of this force in kN?

Hints

- *See Figure 2.2.13 for internal forces and reactions.*

Question 7

Warren Truss

Question 7

Warren Truss

Question 8

Tension Fabric Roof - PTFE

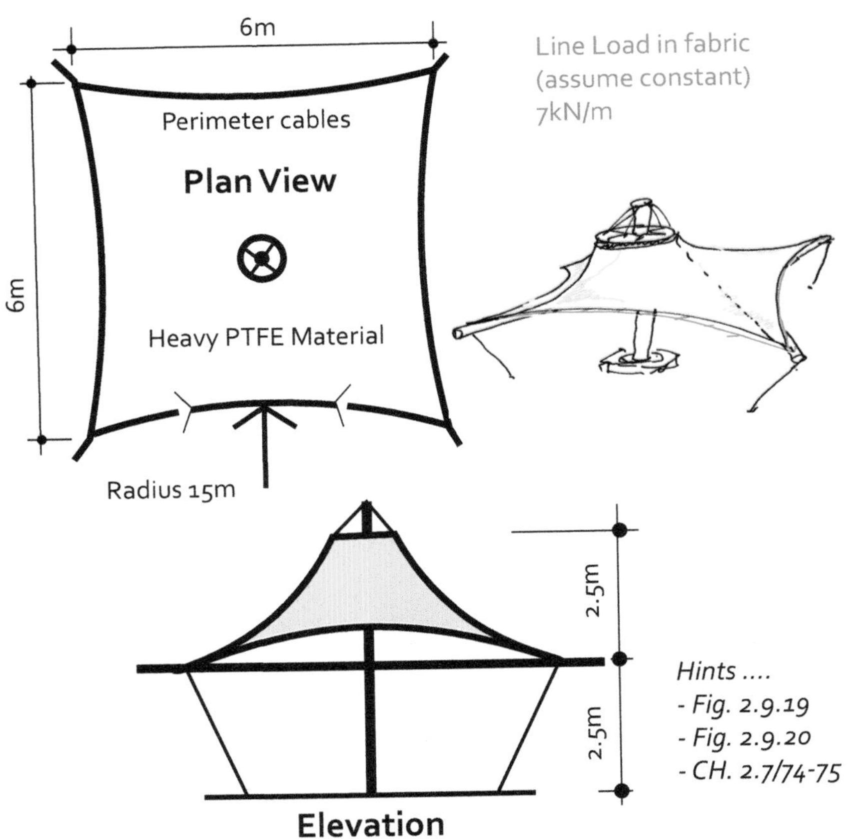

Background

A council is refurbishing a public square and has commissioned a design for a simple canopy to shade stage activities. It has a rectangular plan (6mx6m) and 5m high post with four horizontal struts (booms) projecting out from the post. The fabric has a reinforcing cable on its four edges which is anchored at the boom ends. Corner ties hold down the ends of the booms to foundations.

Question 8

Tension Fabric Roof - PTFE

Questions

8(a)

Draw out a plan of the structure to scale (CAD or by hand) with the 15m radius curve. Now work out the cable tension in kN.

8(b)

Based on a safety factor of 2.5 what is the cable size minimum diameter recommended for the cable load?

Use Macalloy brochure for galvanised strand cables (single strand). *It may be readily found on the internet..or follow* http://www.macalloy.com/brochures-other/tensoteci-cable-systems.

8(c)

On the plan previously drawn trace alongside the curved profiles a reduced radius curve of 7.5m. Find the cable tension in kN. What is the minimum size macalloy cable for this tension load? How do the 7.5m radius and 15m radius compare *aesthetically* on plan?

8(d)

Are the external reactions on the structure more or less (or the same) for the 7.5m radius cable compared with the 15m radius cable? Calculate for both radii the boom compression forces based on the plan geometry. *Booms are the horizontal beams radiating from the centre post.*

Hints

- *See Figure 2.9.19-20 for cable force. The cable is in tension - i.e. 2.9.19 is reversed.*
- *See Chapter 2.7/74-75 for resolution of forces.*

Question 8

Tension Fabric Roof - PTFE

Question 8

Tension Fabric Roof - PTFE

Question 9

Strut Compression

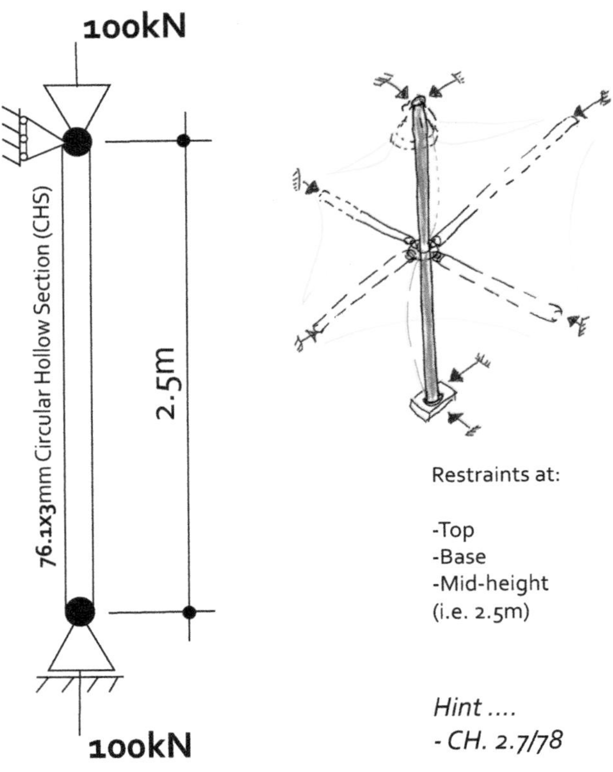

Restraints at:

-Top
-Base
-Mid-height
(i.e. 2.5m)

Hint ....
- CH. 2.7/78

Background

An architect, for the tension roof in Question 8, desires a slender aluminium post for aesthetic reasons for the main mast (post) which has a vertical load of 100kN based on resolved forces. The desired section is a 76.1x3 CHS (circular hollow section).

Question 9

Strut Compression

Questions

9(a)

Find the parameters - section area (cm^2) and moment of inertia (cm^4) and radius of gyration (r_y) for the section. Find the slenderness assuming the post is *restrained* at half height as for the diagram. Is this very slender?

9(b)

What is the Euler critical load in kN?

Using assumed E (modulus) of 70×10^3 N/mm^2 (or mPa).

9(c)

How does this compare with the applied load?

Assume for the purposes of the calculation that this is a factored (ultimate) load.

Hints

- *See Figure 2.7.23 for elevation of post*
- *Aside Chapter 2.7/77-78 for discussion on Euler.*

Question 9

Strut Compression

Question 9

Strut Compression

Question 10

Strut Compression - Larger Section

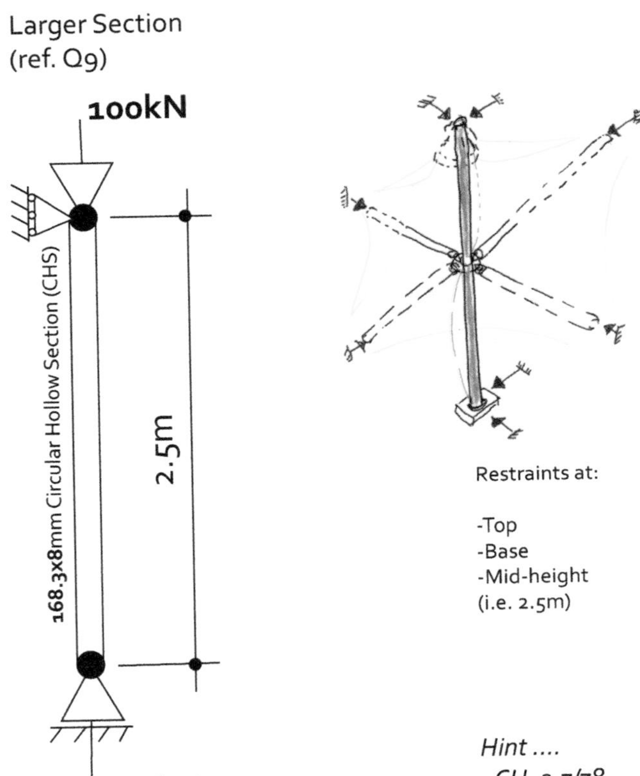

Background

From the previous two questions (8 and 9) the engineer proposes a stockier (less slender) post in aluminium.

Strut Compression - Larger Section

Questions

10(a)

Find the parameters - section area (cm^2), moment of inertia (cm^4) and radius of gyration(r_y in cm) for the section. Find the slenderness assuming the post is restrained at half height as for the diagram. What is the slenderness?

10(b)

What is the Euler critical load in kN? Using assumed E (modulus) of 70x10³ N/ mm² (or mPa).

10(c)

What is the compression stress in the section?

Assume for the purposes of the calculation that this is a factored (ultimate) load.

10(d)

Taking the minimum of the two answers to 10(b) and 10(c), which check is more critical: *Euler or compression stress*?

Is this an appropriate section at this concept design stage?

Hints

- *See Figure 2.7.23 for elevation of post*
- *Aside Chapter 2.7/77-78 for discussion on Euler.*

Question 10

Strut Compression - Larger Section

Question 10

Strut Compression - Larger Section

Question 11

Church Vault Problem

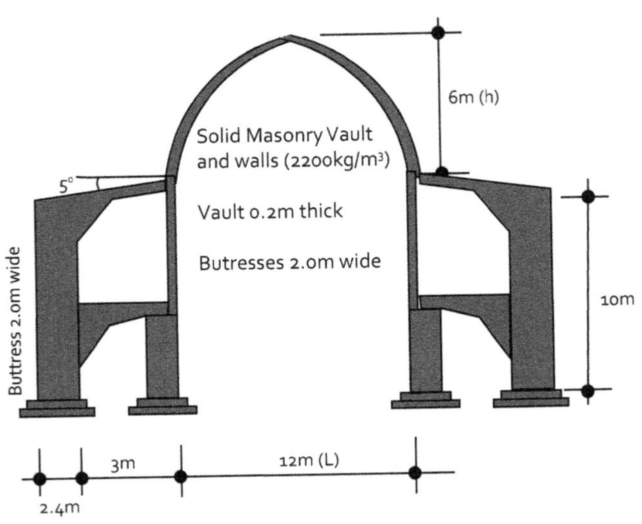

Buttress spacing = 6m

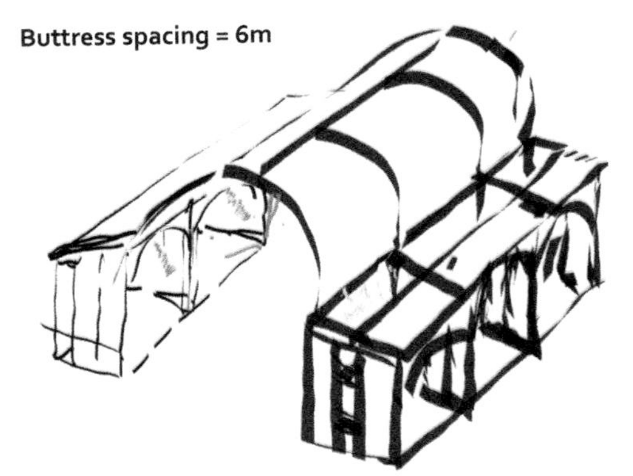

Hints ....

-Assume vault parabolic
-Fig. 2.8.7 (thrust)
- 100kg = 1kN
- Fig. 2.8.17

Background

The main aisle of a church with vaulted roof is built in 6m long bays with Gothic vaulting. Each bay is restrained by a flying buttress (2.4mx2.0m by 10m high). The vault span is 12m with a 6m rise. The vault may be assumed to be built in stone masonry (2200kg/m³) and with a constant thickness of 0.2m (200mm). The buttress is built from similar material.

Church Vault Problem

Questions

11(a)

Investigate using approximate methods the effectiveness of the buttress in restraining the outward thrust of the vault without excessive eccentric loading at the foundation.

Note, in the book (figure 2.8.17), this method is a 'first approximation'. Should the stabilising buttress be checked with respect to the internal stress distribution within the stabilising buttress, it may be demonstrated that theoretically tension may be generated within the internal face. On page 92 various mitigating factors are outlined which in reality show that most buttresses have no tensile stresses, or might allow redistribution within. This is purely a stability check.

Hints

- *See Figure 2.8.7 for horizontal reaction formula*
- *See Figure 2.8.17 for methodology.*
- *Check that the resolved thrust is within the middle third of the buttress/foundation junction.*

Question 11

Church Vault Problem

Question 11

Church Vault Problem

Question 12

Canopy Torsion Problem

Cantilever Section Tapered (steel)

From 200mm deep x 12mm thick (support)
to 100mm deep x 12mm thick.

Tip: assume average depth to work out properties, based on 150 deep (average)

$I_{xx} = 337cm^4$
E (steel) = 205×10^3 N/mm^2
G (steel) = 78×10^3 N/mm^2

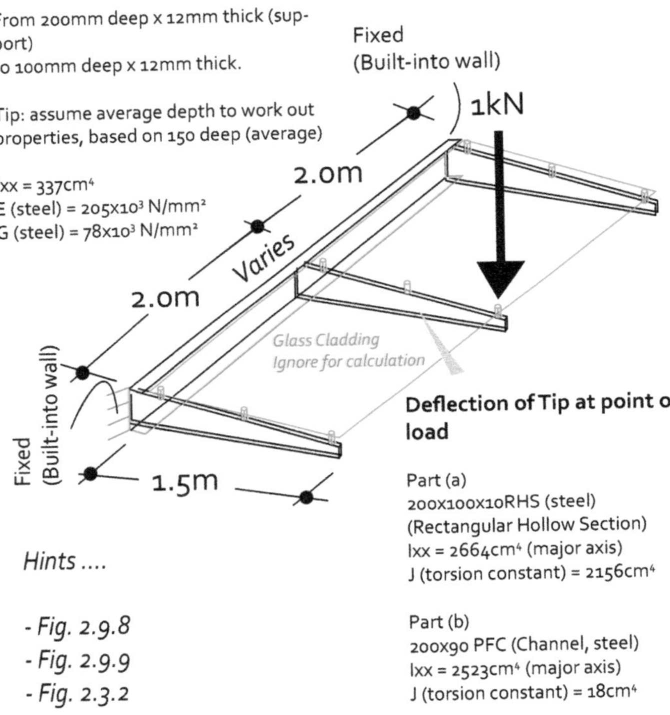

Hints ....

- *Fig. 2.9.8*
- *Fig. 2.9.9*
- *Fig. 2.3.2*

Deflection of Tip at point of load

Part (a)
200x100x10RHS (steel)
(Rectangular Hollow Section)
$I_{xx} = 2664cm^4$ (major axis)
J (torsion constant) = $2156cm^4$

Part (b)
200x90 PFC (Channel, steel)
$I_{xx} = 2523cm^4$ (major axis)
J (torsion constant) = $18cm^4$

Background

A glass canopy in galvanised steel is 4m wide with three tapering arms made from galvanised steel (12mm wide; tapering from 200mm deep to 100mm deep). The architect would like an exposed channel to support the middle arm but will consider a tubular section. The middle arm has to be designed to support a 1kN load (100kg) to support maintenance and cleaning operations.

Canopy Torsion Problem

Questions

12(a)

Find the rotation of a 200x100x20 RHS (*rectangular hollow section*) for the 1kN load.

See diagram for relevant properties.

RHS built into wall at ends.

12(b)

Find the rotation of the architect's *preferred* PFC (*parallel flanged channel*) for the 1kN load.

See diagram for relevant properties.

PFC built into wall at ends.

12(c)

Find the approximate tip deflection of the central cantilever arm for the 1kN load for both sections in 12(a) and 12(b).

Ignore flexural cantilever deflection of the tapered arm.

Would it be an acceptable option to use the architect's preferred PFC section?

Hints

- *See Chapter 2.9/104-105; Figure 2.9.8 and 2.9.9 for methodology.*

Question 12

Canopy Torsion Problem

Question 12

Canopy Torsion Problem

Question 13

Canopy Torsion - Alternative Design

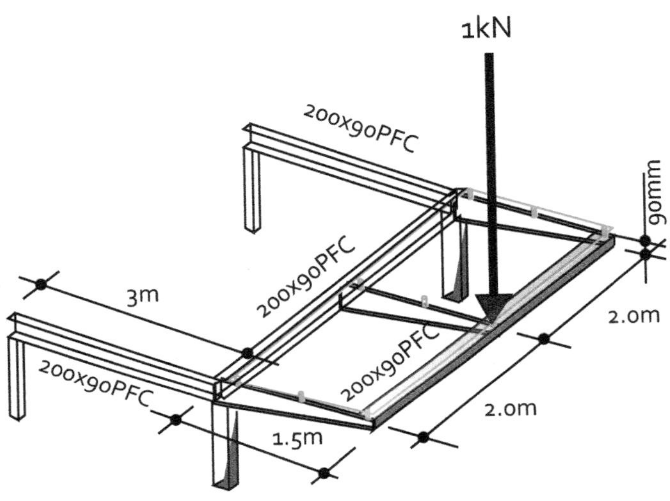

Canopy from Previous Example

Architect decides on an edge channel 200x90PFC on its side (toes down)

200x90 PFC is kept from previous example as desired by architect

Backspanning 200x90 PFCs are added to assist with cantilever

Ixx = 2523cm⁴ (major axis)
Iyy = 314cm⁴ (minor axis)

Hints ....

- *Fig. 2.9.8*
- *Fig. 2.9.9*
- *Fig. 2.3.2*
- *Fig. 2.3.5*
- *Fig. 2.2.3*
- *CH. 2.9/106 aside*

Background

The architect has revised the design for the glass canopy (Question 12) in the hope that the 200x90PFC (support channel over entrance) may be adopted. A 200x90PFC section on its side with flanges pointing downwards, placed at the end of the cantilever arms, is proposed. Holes are also to be cut in the rear cross walls with 200x90PFCs over the openings.

Canopy Torsion - Alternative Design

Questions

13(a)

Find deflection under the 1kN design load in mm for the 200x90 PFC under the cantilever tips assuming simple supports and using the properties on the diagram.

13(b)

Find approximate deflection of the tips of the cantilever arms *which are supporting the new 'toes down' channel in 13(a). Note backspan to be taken into account- it may be assumed that the backspan is same as cantilever average section (conservative assumption).*

13(c)

What is the aggregate (*sum of deflections*) for the two calculations 13(a) and 13(b) above?

13(d)

How does the stiffness of the option 200x90PFC in torsion in question 12(b) compare with the stiffness of the in question 13(a)?

13(e)

Is it appropriate to adopt the fascia (edge) channel over the entrance as the architect desired in Question 12 with this structural arrangement?

Hints

- *See Figures 2.3.2 and 2.3.5.*
- *See Figure 2.2.3*
- *See Aside Chapter 2.9/106.*

Question 13

Canopy Torsion - Alternative Design

Question 13

Canopy Torsion - Alternative Design

Question 14

Deep Beam

Reinforced Concrete Wall Spanning over retail area

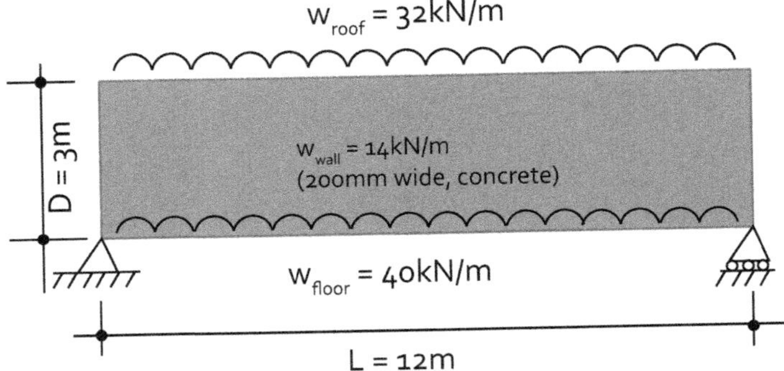

Loading comprises Dead and Live combined

For Ultimate loads apply safety factor 1.50

For deflections use original loads without safety factor

Concrete Strength (cube/cylinder)
Assumed to be 35 N/mm² *or* mPa

Hints ....

- *Fig.* 2.11.2

Background

A reinforced concrete restaurant building has a rear wall which has to span 12m onto columns over the main glazed entrance. This will serve to support a flat slab floor and roof. The wall is 0.2m (200mm) thick. It is not certain at this stage in the design whether the roof or floor slab may be used structurally as flanges but they will *provide lateral restraint along the wall.*

Deep Beam

Questions

14(a)

Find the maximum bending moments in the wall and shear forces at the supporting columns using a 1.5 factor as noted in the diagram.

14(b)

Using a deep beam approximation find an appropriate number of 25mm diameter reinforcement bars with a yield strength of 500N/mm^2 . Round up to an even number. *Assume concrete strength circa 35N/mm^2.*

14(c)

Find shear stress in the section at the worst position. Is this high or low? With 10mm diameter shear links (vertical rods in a loop around the section) what sort of spacing is reasonable (usually around 150mm-250mm centres)?

Note: question 14(c) not covered in book.

Hints

- *See Figure 2.11.12.*
- *Shear steel (Area = Asv) approximate calculation...see below*
- *Approximate area steel bar, for shear (Asv) = Sv.w.Vs/fy, where Sv is spacing of shear links, Vs is shear stress, w - width section, fy - reinforcement yield*

Question 14

Deep Beam

Question 14

Deep Beam

Question 15

Deep Beam - with Openings

Reinforced Concrete Wall Spanning over retail area with Openings

Loading from previous exercise (deep beam)
w_{roof} = 32kN/m, w_{floor} = 40kN/m, w_{wall} = 14kN/m

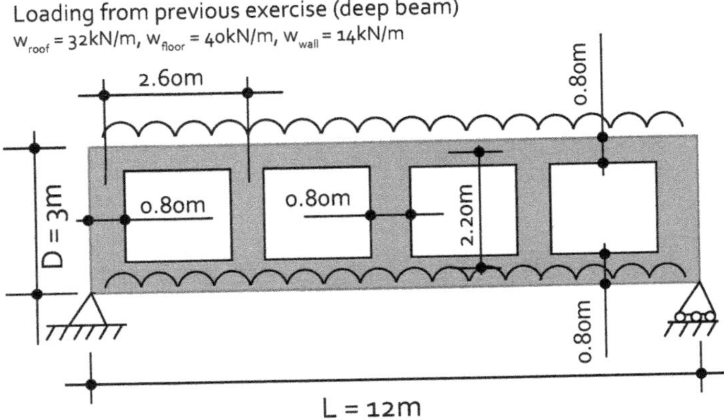

Loading comprises Dead and Live combined

For Ultimate loads apply safety factor 1.50

For deflections use original loads without safety factor

Concrete Strength (cube/cylinder)
Assumed to be 35 N/mm² *or* mPa

Hints ....

- *Fig.* 2.6.22
- *Fig.* 2.11.2

Background

In the restaurant design from Question 14 the owner would now like openings in the wall. This becomes a different solution since the wall is now much closer to a vierendeel beam than deep beam in behaviour.

Deep Beam - with Openings

Questions

15(a)

As a vierendeel analysis for first approximation find the worst case bending moments in the 800 deep vertical and horizontal beam elements.

15(b)

Find the midspan bending moment for the vierendeel *overall*. Using this what is the estimated tension and compression in the horizontal beam elements on the side of the central pillar?

15(c)

If the long term effective modulus of the reinforced section is 12×10^3 N/mm^2 (mPa) find an approximate deflection under the unfactored loading from wall, floor and roof based solely on flexural beam theory.

Compare with a 2d computer model of the frame.

15(d) - *Note question 15(d) is not covered in the book.*

Using bending lever arm of 540mm in the section find an appropriate reinforcement arrangement in the sections taking the worst case bending.

Check this for the support horizontal elements only. Concrete strength 35N/mm^2.

What sort of shear links or bars would be needed? Assume all shear reinforcement has a yield strength of 500N/mm^2 and use ultimate load factor of 1.50 to assess internal forces.

Hints

- *See Figures 2.6.22.*
- *See Figures 2.11.2.*

Question 15

Deep Beam - with Openings

Question 15

Deep Beam - with Openings

Answers to Question 1

Simply Supported Beam

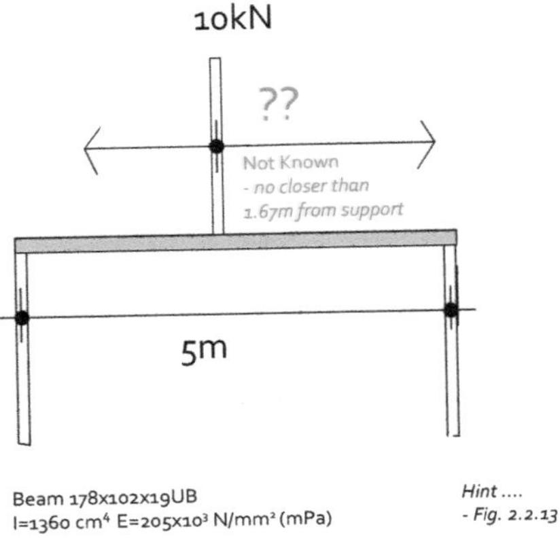

Beam 178x102x19UB
I=1360 cm^4 E=205x10^3 N/mm^2 (mPa)

Hint ....
- Fig. 2.2.13

Answers

1(a)

12.5kNm - worst case is to assume the point load is in midspan position (M=PL/4). If the load was assumed to be in third span position the maximum bending moment would be 11.2kNm (89% of midspan). See figure 2.2.13.

1(b)

6.7kN - worst case shear at support . This was assuming load at third span position (1.67m from support). The opposite support shear is 3.3kN. See figure 2.2.13.

1(c)

9.3mm - Deflection from figure 2.2.3.

1(d)

Yes, it is acceptable. If the deflection is expressed as a ratio of span/ deflection: the ratio is 5000mm/9.3mm = 540 (approximately). Under normal circumstances a combination of dead and live load deflection is acceptable if this ratio is 300 or more.

Simply Supported Beam

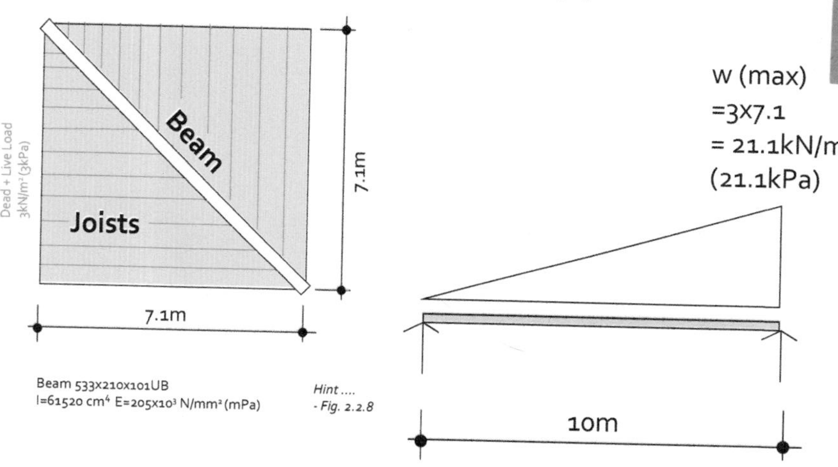

Answers

2(a)

10m span (7/cos45°); peak load 21.1kN/m- See above. Multiply Area load x loaded width.

2(b)

175kNm - This is $wl^2/8$ for a uniformly distributed line load (UDL) of 14kN/m; obtained from the peak triangular load. (21.1x2/3) with reference to figure 2.2.8.

2(c)

14.5mm - Deflection from figure 2.2.3 [$5wl^4/(384EI)$]. UDL from 2(b) above of 14kN/m.

2(d)

70kN for both cases. Maximum shear for the triangular loading is under the support on the right hand side (RHS). Using the equivalent UDL will always give a safe shear/support reaction.

Answers to Question 3

Cantilever Beam and Support Post

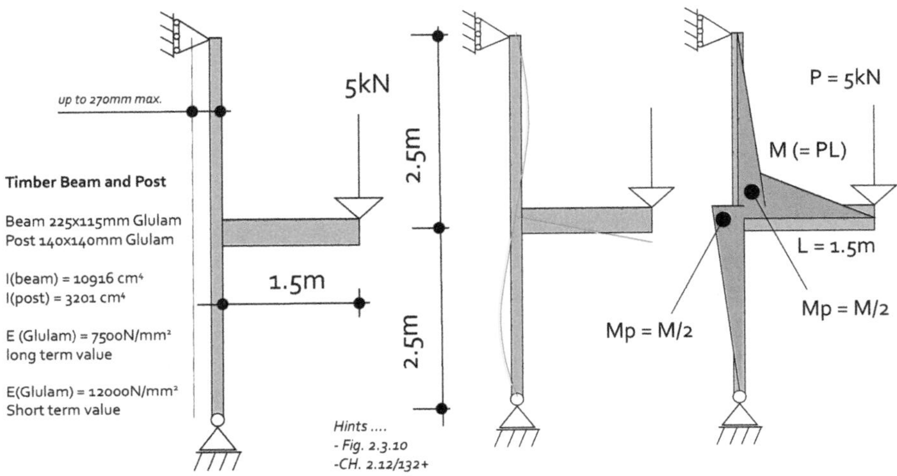

Answers

Spreadsheet Available
www. struartapp.com/resources/figure 2.3.11

3(a)

7mm - from figure 2.3.2 deflection is $PL^3/3EI$ for an encastre support (fixed).

See diagrams above (RHS). The deflected shape induces curvature in the posts. Bending moment at end of beam is PL and post (*centre*) PL/2.

3(b)

19.5mm - Rotation in each post at the beam support $[M_p h/(3EI)]$ where M_p is the post bending; h=2.5m; M_p=M/2=3.75kNm). Then multiply by beam span (0.013 radiansx1500mm).Method: CH. 2.9/105 solution 2 for balustrade.

3(c)

No, it is excessive@L/57 ratio. Deflection (short term E) gives natural frequency circa 4Hz -ref. CH. 2.12/132) - *use short term E for natural frequency*.

3(d)

Marginal - 10mm long term/6mm short term - f_n= 6.5Hz. If loading can be reduced on *refined* design then these sizes may be found to be acceptable.

Continuous Beam

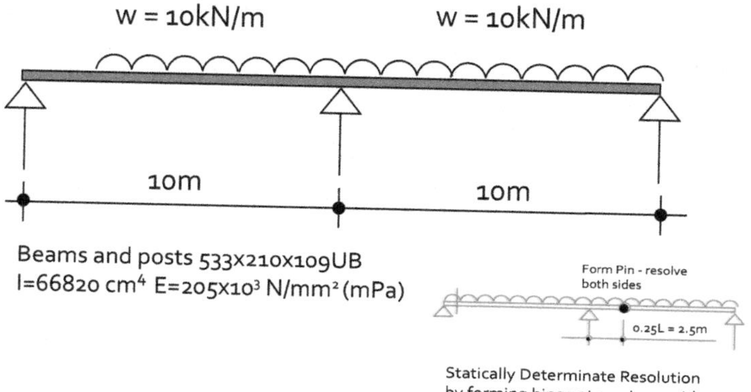

Hints ....
- Fig. 2.4.1
- Fig. 2.4.7
- Fig. 2.3.5
- Fig 2.3.7

Answers

4(a)

125kNm (Support, hogging); 70.3kNm (maximum span moment); ***R_{centre}=125kN - support reaction***.

4(b)

4.25mm - Can be found by finding the shear in the hinge and adding together the cantilever with backspan from figure 2.3.5 and cantilever with UDL from figure 2.3.7 (*factor pro-rata finding pure cantilever with relevant diagrams*). Add this to the right hand side span (0.75L) from figure 2.2.3. Hint: UDL creates uplift deflection (-ve) at hinge cantilever tip (-5.25mm).

This is a time consuming exercise but will give a good appreciation of using superposition to estimate deflections.

4(c)

125kNm (Support, hogging); 70.3kNm (maximum span moment); 3.8m deflection ($wL^4/(185EI)$). The results for 4(a) and 4(c) are the same (*within 10% deflection)*. A computer model gave 4.2mm max.

Answers to Question 5

Three Storey Wind Frame

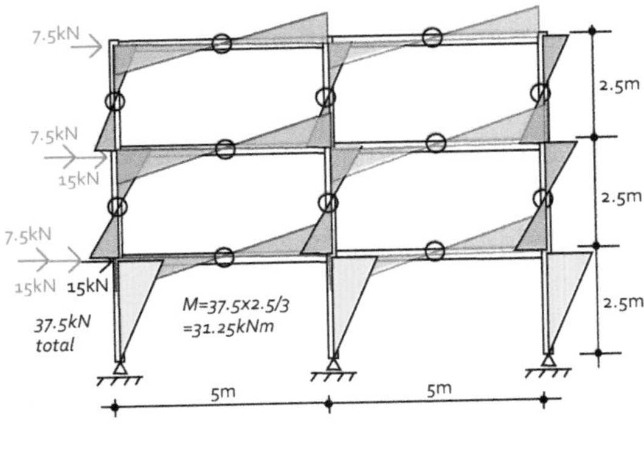

Beams and posts 254x146x31UB
I=4413 cm⁴ E=205x10³ N/mm² (mPa)

Plastic Modulus (S) = 393 cm³

Hints ....
- *Fig. CH. 2.5/43-44*
- *Fig. 2.2.3*
- *CH. 2.12/132-133*
- *Ch. 2.12/138-139*

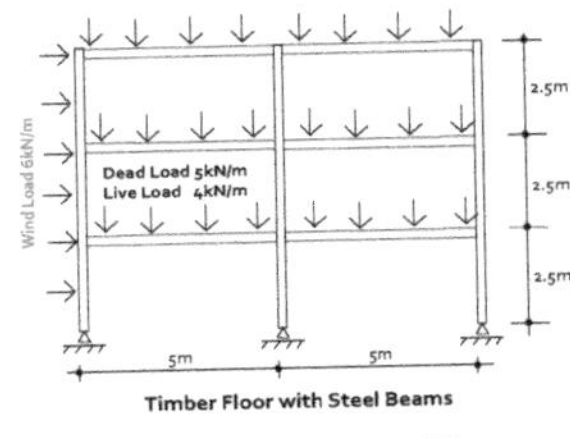

Beams and posts 254x146x31UB
I=4413 cm⁴ E=205x10³ N/mm² (mPa)
Plastic Modulus (S) = 393 cm³

Hints ....
- *Fig. CH. 2.5/43-44*
- *Fig. 2.2.3*
- *CH. 2.12/132-133*
- *Ch. 2.12/138-139*

Answers

5(a)

42kNm=Ultimate Moment; capacity from plastic modulus=108kNm > 42. Sufficient for fully laterally restrained beam with simply supported moment.

5(b)

Wind diagram - see above. Approximate moment in posts - 31kNm.

5(c)

19.8mm (approximate) - 10.8mm FF, 5.4mm 2F, 3.6mm Roof. Found by *deflection* = wh³/(2EI) for first floor (w=37.5kN/3 per post), then adding 50% and 33% approximate upper storey deflections laterally. See Ch. 1.2/45 in book. *A computer model gave 19.3mm.*

5(d)

5.4mm *(Dead+10%Live load 5.4kN/m)*. Fundamental natural frequency from Ch. 2.12/132 is *17.7/sqrt.(deflection)* = ***8Hz***. 8Hz with timber floors ***may be a problem:*** *combined joist and steel beam natural frequency will be less than 8Hz from the aside on page 138*. See Ch.2.12/134 and 138.

Warren Truss

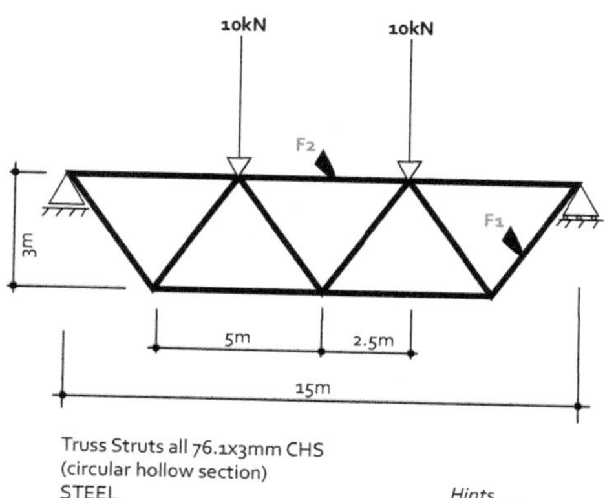

Truss Struts all 76.1x3mm CHS
(circular hollow section)
STEEL
$E=205\times10^3$ N/mm^2; $G=78\times10^3$ N/mm^2

Area of section = 6.9cm^2

Hints ....
- Fig. 2.2.9
- CH. 2.6/53-55

Note: for any compression elements buckling checks should be carried out.

In the case of 6(c) - F2
EC3 design chart showed ultimate capacity of 76.1x3CHS (Le=5m)
= 35kN (approx.)

Or by Euler ref. Ch. 2.7/78 = 37kN

Applied load (ultimate)
=17kNx1.50= 25kN < 35

Answers

6(a)

F1 = 13kN - Tension. Resolved force is 10kN/cos39°. Note a *true* Warren Truss comprises equilateral triangles (60° angles internally) - but it is very similar.

6(b)

F2=17kN - Compression. Strut compression from figure 2.6.7 in centre is 10x5/3=50kNm/3=17kN. Note for computer model check use *sliding* support.

6(c)

28N/mm^2 - F1; 37N/mm^2 - F2. ***Buckling: see note above.*** The truss may be sized for all elements by finding these two strut loads. Note buckling checks are needed between restraints for the top chord (F2) in compression.

6(d)

2mm in flexure; 0.5mm in shear - total 2.5mm. See Ch. 2.6/54-55 in book. Virtual I (inertia) of section 306×10^3 cm^4; area 13.8 cm^2; flexural deflection $23wL^3/(648EI)$ figure 2.2.9; shear deflection $wL^2a/(8AG)$ Ch. 2.6/55; a=6/5 *where a is notation for alpha in book. Computer model gave circa 2.7mm.*

Answers to Question 7

Warren Truss

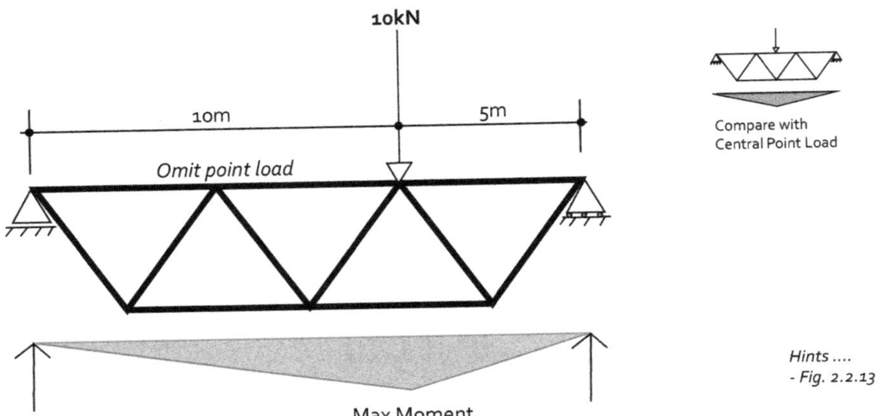

Hints ....
- Fig. 2.2.13

Answers

7(a)

3.3kN shear right hand side (RHS) support; 6.7kN shear LHS; 33.3kNm maximum bending moment.

7(b)

37.5kNm. See figure 2.2.13 load in middle third. Maximum bending moment in 7(a) is 33.3kNm, which is 89% of that of beam with central point load, same span ($PL/4$).

7(c)

8.7kN - Tension. The reaction at the support is 6.7kN maximum. Tension in strut at support (F1) is 6.7/cos39°=8.7kN. *Note this is 66% of the load in the strut from question 6(a); halfing the load does not half the reaction in this case where the load is nearer one support than the other.*

Tension Fabric Roof - PTFE

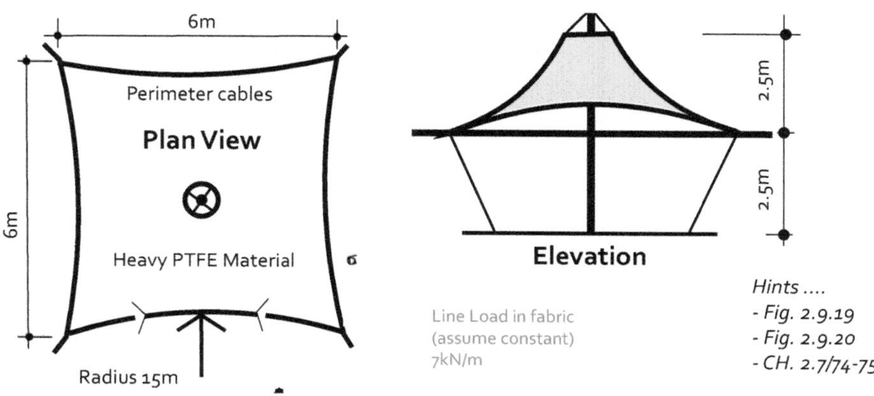

Answers

8(a)

105kN - tension in cable 15m radius. From reverse of Figure 1.9.19 T=PR.

8(b)

20mm cable single strand - *380kN - minimum breaking load of 20mm open strand galvanised - from Macalloy link*. Multiply the load in the cable by 2.5 (safety factor) = 263kN - nearest cable size which is sufficient - 20mm cable.

8(c)

52.5kN - tension in 7.5m radius. Loadx2.5 (safety factor)= 131kN - nearest cable size 12mm - minimum breaking load 135kN.

By observation with similar structures this curve looks to be more in proportion. Thinner cables mean neater/smaller fittings. Drawing to scale is important for visual reference.

8(d)

No change in external reactions. For R=15m: boom compression **~ 178kN**, and for R=7.5m: boom compression **~ 95kN**.

Answers to Question 9

Strut/Mast Compression

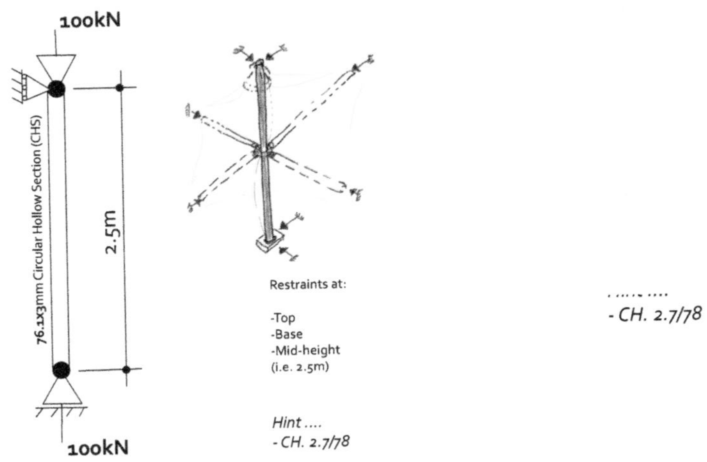

- CH. 2.7/78

Answers

9(a)

6.9 cm² (6900 mm²) area; 46.1 cm⁴ - moment of inertia; r_y=2.6cm - radius of gyration; Le = 2.5m (1.0L); Slenderness (Le/r_y) = 96. The radius of gyration is not explicitly mentioned in the book. It is found as *sqrt(I/A)* in cm.

*Slenderness of steel strut 200-250 is in the very slender range. A slenderness of **96** is not particularly high.*

9(b)

50958N or 51kN. From Euler page 78 - $P_{cr} = \pi^2 EI/(L_e^{\ 2})$. It is perfectly valid to use strut buckling tables from EC3 (Eurocode 3) for instance, but Euler for reasonably slender sections is close in value (*for example see note on answers to question 6 where Euler value is 5% greater than EC3 value)*.

If using Euler for low slenderness check the *compression stress* against yield.

9(c)

It is less than the applied load (100kN). The section will have to be increased in size in order to carry the load.

Strut/Mast Compression

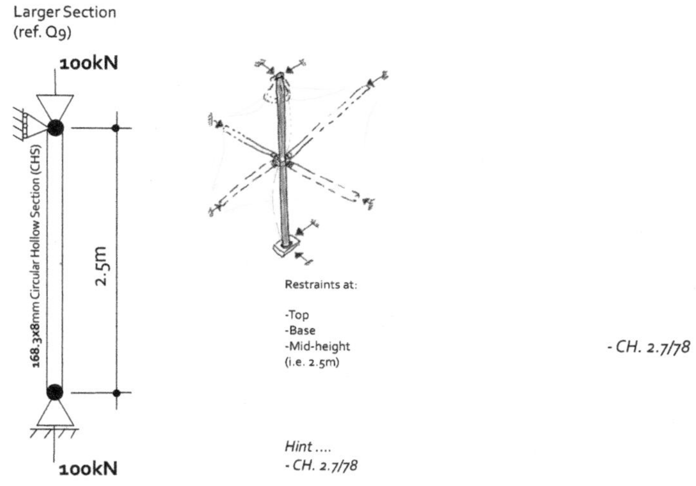

Answers

10(a)

40.3 cm^2 (4030 mm^2) area; 1297 cm^4 - moment of inertia; r_y=5.7cm - radius of gyration; Le = 2.5m (1.0L); Slenderness (Le/r_y) = 44. The radius of gyration is not explicitly mentioned in the book. It is found as *sqrt(I/A)* in cm.

10(b)

1430x10^{3}N or 1430kN. >>100kN

10(c)

24N/mm^2 (mPa). Yield stress for aluminium varies but can be assumed to be approximately 100mPa. The section is within capacity.

10(d)

355N/mm^2 is equivalent Euler stress (or convert applied compression stress to equivalent load, multiplying by area). 355>>24. *For low slenderness columns/struts Euler is not necessarily the critical check. This should always be borne in mind for this approximate method.*

Answers to Question 11

Church Vault Problem

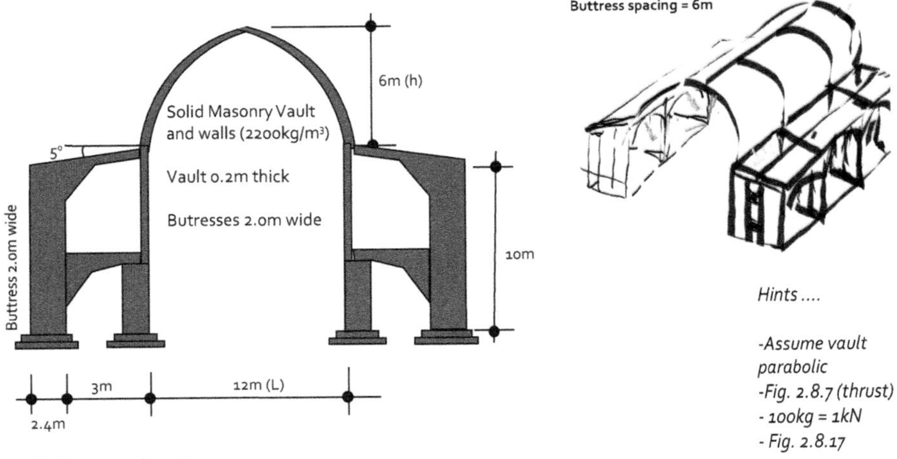

Hints ....

-Assume vault parabolic
-Fig. 2.8.7 (thrust)
- 100kg = 1kN
- Fig. 2.8.17

Answers

11(a)

1056kN (105 tonnes) weight of buttress; 79kN lateral vault thrust; thrust is within middle third, is stable based on first approximation.

The methodology in **figure 2.8.17** may be followed.

The lateral or sideways thrust from the vault is found from $R_h = wL^2/8H$ = 22kPax6m(bay)x0.2mx12^2/8/6 = **79kN**.

Buttress weight = 22x2.0(wide)x10m(high)x2.4m= **1056kN**.

Thrust location = 79kN(thrust)x10/1056 = **0.79m** < 0.80

Therefore it is within the middle third of the buttress, further alleviated by the spread base.

This may easily be drawn to scale as a vector of forces with resultant.

Canopy Torsion Problem

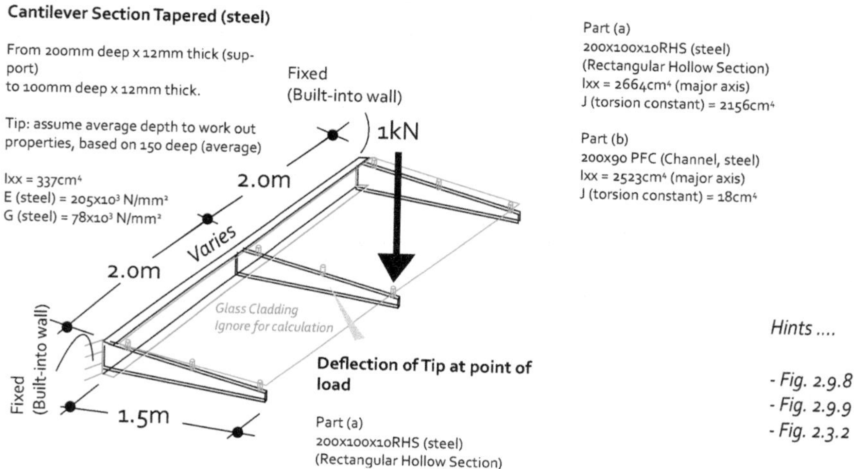

Answers

12(a)

9×10^{-4} radians. See pages 102 in the book for rotation formula,

Rotation = TL/GJ, where T in this case is 50% of the applied torsion, and L is 50% of the span of the beam in question. For steel shear modulus = 78×10^3 N/mm² (mPA).

Rotation = 1.5kNm$\times 10^3 \times (4000\text{mm}/2)/78 \times 10^3/2156 \times 10^4 = 9 \times 10^{-4}$ radians.

12(b)

0.106 radians. Using method as above. This is approximately 6 degrees.

12(c)

1.3mm tip deflection for the RHS section; 158mm for the PFC section. Multiply rotation by the cantilever length (1.5m) for deflection. e.g. 0.105x1500mm= 158mm for the PFC section. This is high (a ratio of 1/10 for the cantilever span).

The drop in the tip would not be acceptable on this basis.

Answers to Question 13

Canopy Torsion - Revised Design

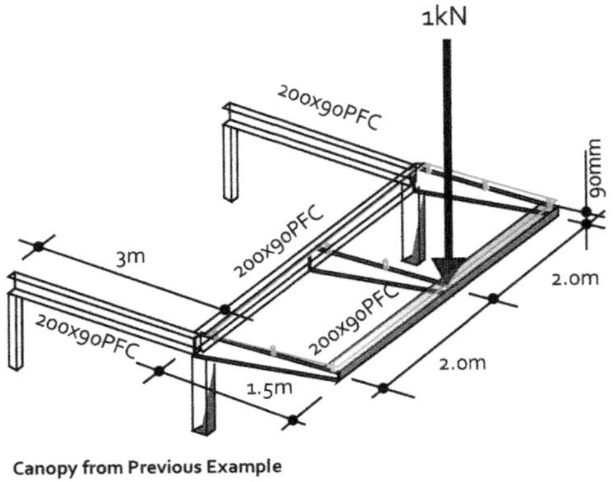

Canopy from Previous Example

Canopy from Previous Example

Architect decides on an edge channel 200x90PFC on its side (toes down)

200x90 PFC is kept from previous example as desired by architect

Backspanning 200x90 PFCs are added to assist with cantilever

Ixx = 2523cm⁴ (major axis)
Iyy = 314cm⁴ (minor axis)

Hints ....

- *Fig. 2.9.8*
- *Fig. 2.9.9*
- *Fig. 2.3.2*
- *Fig. 2.3.5*
- *Fig. 2.2.3*
- *CH. 2.9/106 aside*

Answers

13(a)

2mm. Figure 2.2.3, simply supported beam with central point load deflection = $PL^3/(48EI)$ - where P=1kN; L=4m; E=205×10^3 N/mm^2; I=314cm^3 - minor axis.

13(b)

2.3mm worst case. See figure2.3.5. I value for the tapered section average depth (150mm) is $150^3 \times 12/12 \times 10^4$ = 337.5cm^3. Backspan PFC (I) 2523 cm^3 - *first approximation* 337.5 cm^3 used for backspan. Pure cantilever 0.81mm ($PL^3/(3EI)$). Pro-rata for '2L' backspan *figure 2.3.5* = 8.5x0.81/3 = 2.3mm.

13(c)

4.3mm. 13(a)+13(b).

13(d)

Option 13(c) is circa 37 times stiffer than option 12(b).

13(e)

Yes this is much more rigid. The PFC fascia section is now an option.

Deep Beam

Reinforced Concrete Wall Spanning over retail area

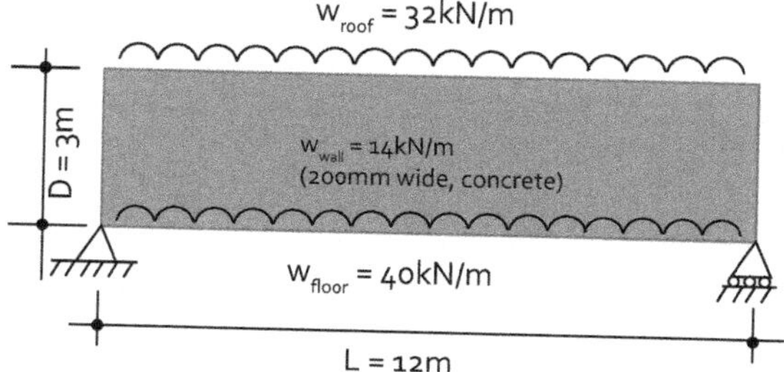

Hints ....

- Fig. 2.11.2

Loading comprises Dead and Live combined

For Ultimate loads apply safety factor 1.50

For deflections use original loads without safety factor

Answers

14(a)

2322kNm Maximum factored bending moment - centre. 774kN factored shear - support. Based on figure 2.2.1 and 2.2.2 formulae. Total ultimate line load = (40+32+14)x1.5=129kN/m.

14(b)

6 reinforcement bars. Figure 2.11.12. Lever arm = 0.62L=0.62x3m=1.86m. Tension (bars) = M/La = 2322/1.86 = 1248kN. Area steel = 1248×10^3N/500N/mm^2 = 2496 mm^2. Area 25mm bar = 490 mm^2. 2496/490 = 5.1 bars - round up to 6.

14(c)

1.3N/mm^2 (mPa); medium value Max approx. 5N/mm^2. 250mm spacing .

Approximate area steel bar (A_{SV}) = S_v.w.V_s/f_y, where S_v is spacing of shear links, V_s is shear stress, w - width section, f_y - reinforcement yield. A_{sv}=250x200x1.3/500=130mm^2 (Area bent leg 2 no. 10mm bars is 157mm^2>130). *This is not covered in the book* but is useful estimation of shear reinforcement.

Answers to Question 15

Deep Beam - with Openings

Reinforced Concrete Wall Spanning over retail area with Openings

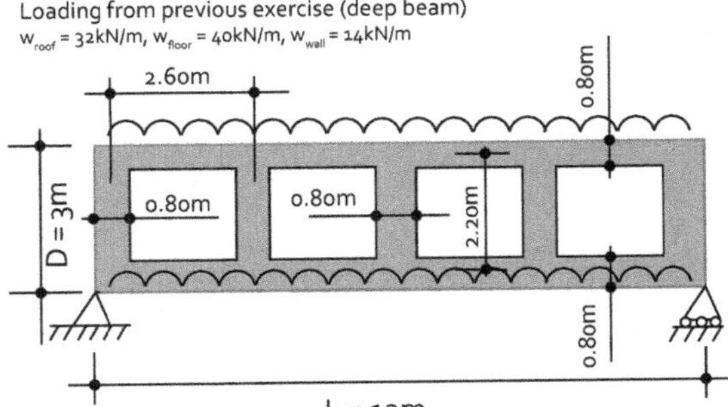

Hints ....

- Fig. 2.6.22
- Fig. 2.11.2

Loading comprises Dead and Live combined

For Ultimate loads apply safety factor 1.50

For deflections use original loads without safety factor

Answers

15(a)

503kNm. Figure 2.6.22. Shear at end from 14(a) 774kN.
M = shear x H/4 = 774x2.6/4 = 503kNm - the hole boundary is approximately 2.6m (=3-0.0.4) using the longer dimension (not 2.2m).

15(b)

1055kN. See figure2.6.23. Midspan moment from 14(a) = 2322kNm. T/C top and bottom = M/D = 2322/(3-0.8) = 1055kN. *A computer model gave T/C = 914kN and moment max. 515kNm..*

15(c)

5mm. I section - from Figure 2.6.10 = $2x(2200/2)^2x800x200/1x10^4 = 39x10^6$ cm^4. Deflection, = $5wL^4/(384EI)$; $E=12x10^3$ N/mm^2, as given. *A computer model gave circa 18mm.Beam theory is unreliable - behaves more as frame.*

15(d)

Not an essential question as it is not covered in the book. See next page for possible answer with discussion.

Deep Beam -with Openings

Notes

There is no 'right answer' to this. The assessment of reinforcement was a first approximation. Points of note:

Overall shear stress at support circa 2.6N/mm². A check to a British code indicated shear reinforcement could be reduced, taking into account the inherent shear capacity of the concrete.

It also indicated an overall area of reinforcement 406kg/m³ which is high. A more economical design would utilise the roof slab and floor slab as flanges (like a channel section).

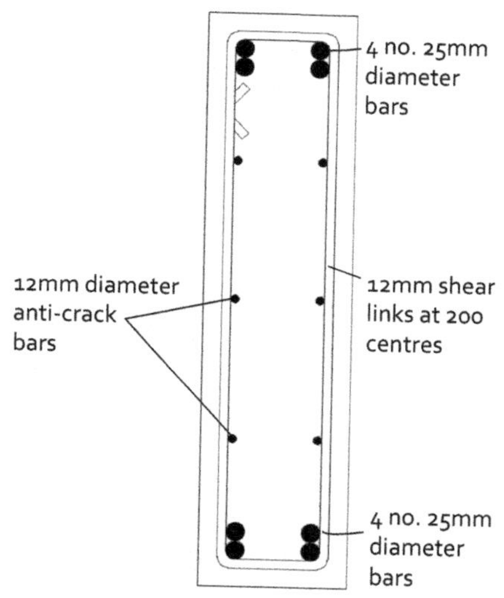

Answers

15(d)

See Section above with reinforcement bar arrangement.

Finding lever arm: Area of bar in top and bottom was found assuming the lever arm was based on 90% of the top as compression zone (400mmx0.9=360mm).

The centre of tension from bottom is: *30mm cover for bars + 12mm (link) + 25mm bar = 67mm (approximate)*

Total lever arm 800mm-360/2-67= 553mm.

Tension force = M/La=503/0.553=909kN

Area bars needed = 909x10³/500 = 1818mm² - 4no. 25 diameter = 4x490 = 1960mm² > 1818. Put bars in the top and bottom faces as the bending tension zone is top and bottom.

Link area @ 200 spacing (approximate) = (Asv) = S_v.w.V_s/f_y, where Vs = 774E3/200/2/(800-67)= 2.6N/mm². Area Asv =200x200x2.6/500 = 208mm² T12 (legs pairs) area = 113x2 = 226mm² > 208

Lightning Source UK Ltd.
Milton Keynes UK
UKOW06f1336100817
307066UK00009B/96/P

9 781782 224792